Energy Matters

ISSUES

Volume 97

Editor

Craig Donnellan

Educational Publishers
Cambridge

First published by Independence
PO Box 295
Cambridge CB1 3XP
England

British Library Cataloguing in Publication Data
Energy Matters – (Issues Series)
I. Donnellan, Craig II. Series
333.7'9

ISBN 1 86168 305 7

Printed in Great Britain
MWL Print Group Ltd

Typeset by
Claire Boyd

Cover
The illustration on the front cover is by
Simon Kneebone.

CONTENTS

Introduction

Energy Matters is the ninety-seventh volume in the **Issues** series. The aim of this series is to offer up-to-date information about important issues in our world.

Energy Matters looks at renewable energy and energy efficiency.

The information comes from a wide variety of sources and includes:
Government reports and statistics
Newspaper reports and features
Magazine articles and surveys
Website material
Literature from lobby groups
and charitable organisations.

It is hoped that, as you read about the many aspects of the issues explored in this book, you will critically evaluate the information presented. It is important that you decide whether you are being presented with facts or opinions. Does the writer give a biased or an unbiased report? If an opinion is being expressed, do you agree with the writer?

Energy Matters offers a useful starting-point for those who need convenient access to information about the many issues involved. However, it is only a starting-point. At the back of the book is a list of organisations which you may want to contact for further information.

*****

The case for renewable energy

Information from NATTA

Introduction

The winds, waves, tides and sunlight come to us free, and using these natural energy sources does not create any pollution, or make us reliant on resources which are constrained and exhaustible. This short overview argues the case for a switch to using these renewable resources as an alternative to both fossil fuels and nuclear power.

Why we have to change

Until recently it was thought that the main problem in relation to the world's energy resources was that some of them, oil especially, would soon run out. That is still an issue. Nowadays, however, the problem seems to be that we cannot afford to risk using up whatever reserves are left since to do so will irreversibly damage the global climate system.

Burning coal, oil and gas inevitably produces carbon dioxide gas, a gas which plays a major role in climate change. After this gas is emitted from power stations and cars, it travels up to the upper atmosphere where, along with other so-called 'greenhouse gases' like methane, it acts like the glass in a greenhouse, trapping the heat radiation from the sun. The result is that the greenhouse, in this case the earth, heats up. If we carry on burning off fossil fuels, over the next century average global temperatures could rise by 6 degrees Celsius or more. That would change life on earth dramatically. The climate and weather system would be changed so that there would be increased occurrence of very severe storms and floods, along, at other times, with increased droughts. The increase in temperature would melt the ice caps and lead to the thermal expansion of the seas, so that sea

---

Burning coal, oil and gas inevitably produces carbon dioxide gas, a gas which plays a major role in climate change

---

levels could rise by up to a metre. Key low-lying areas would be inundated – including many rice growing areas – and many millions would become homeless.

The social and economic cost of climate change on this sort of scale is incalculable, and would continue to grow year by year. Certainly it would be much more than the cost of halting the use of fossil fuels – large though that would be. At present nearly 80% of the energy we use on earth comes from fossil fuel, so it will be a huge task to move away from that. However, the cost of making this change has to be put in context. At some stage, all the existing power plants will have to be replaced in any case, as they reach the end of their working life. So we would just be replacing them with technologies that did not use fossil fuel. Of course, given the urgency of the climate change issue, we may not be able to wait for old plants to be replaced – we will have to shut some down early. But even so, the total cost of the switch over should

FOSSIL FUEL

RENEWABLE ENERGY

Simon Kneebone

not be too difficult to bear, especially when you realise that, in addition to climate change, burning fossil fuels also leads to other social and environmental costs, for example as a result of acid emissions.

Most countries in the world are trying to respond to the climate change threat and have agreed to cut emissions. In 2003 the UK government decided to try to reduce carbon dioxide emissions by 60% by 2050. How can this be done?

The alternatives to reliance on fossil fuel

The easiest way to reduce reliance on fossil fuels is to generate and use energy more efficiently. In terms of generation, using gas instead of coal in modern combined cycle power plants provides some emission improvements, and Combined Heat and Power (CHP) plants, which use the otherwise wasted heat, can more than double overall energy conversion efficiency.

On the consumption side, at present we waste much of the energy that we get from fossil fuels in poorly designed and insulated houses and in inefficient energy-using devices. In part that is because, until recently, fossil fuels were relatively cheap and plentiful and the impacts of burning them were not appreciated, so efficiency didn't matter too much. Consequently it should be relatively easy to make some large savings – at least initially. A wide range of energy efficient devices are available – fridges, washing machines, lighting systems and so on. However, once all the quick and easy energy savings have been made, the scope for reducing waste will be less. Worse still, it seems likely that overall demand for energy will continue to rise, even given sensible efforts to reduce waste – energy use has risen globally by 1-2% year by year and, with more and more people around the world wanting to copy the West's consumer lifestyle, this growth seems unlikely to slow.

The easiest way to reduce reliance on fossil fuels is to generate and use energy more efficiently

---

Of course you could say that we should all consciously choose to avoid continual growth in material consumption. That would certainly reduce energy growth. But a voluntary austerity/conscious frugality approach has only limited appeal to most people. It may nevertheless be that we can move to more sustainable forms of consumption by adjusting our lifestyles without having to make major sacrifices. That could help, but even so, we would still need energy. As it stands then, it seems that, at best, energy conservation, through the use of more efficient energy systems and adjustments in lifestyles, can perhaps only slow the overall expansion in energy use. So, while energy conservation is obviously vital, we will also need new sources of non-fossil energy.

One option is nuclear power. Nuclear plants supply nearly 7% of the world's energy at present, in the form of about 18% of the world's electricity, and they do not generate carbon dioxide. But they do generate a lot of very dangerous and long-lived radioactive by-products and wastes, some of which can be used to make nuclear weapons and some of which have to be stored for millennia to keep them safely separate from the biosystem. Nuclear power is also expensive, costing significantly more than conventional energy sources, and the power plants and ancillary plants are prone to occasional leaks and accidents, some of which can have major impacts, as the Chernobyl disaster in the Ukraine indicated. As a consequence nuclear power has fallen out of favour in most countries of the world. For example most of Europe is phasing out its use of nuclear power. Instead, many countries are looking to renewable energy as the way ahead.

There are a lot of renewable energy sources – solar, wind, wave, and so on. What they have in common is that, unlike fossil or nuclear fuels, using them creates no emissions or wastes and these energy sources cannot be exhausted. They will be available indefinitely – as long as the planet lasts.

■ The above information is from *The Case for Renewable Energy*, May 2004, £1, produced by NATTA.
© *Network for Alternative Technology and Technology Assessment (NATTA)*

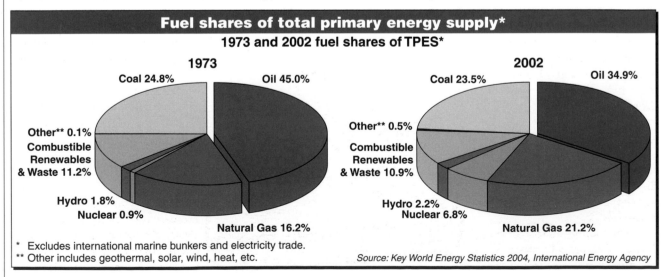

Fuel shares of total primary energy supply*

1973 and 2002 fuel shares of TPES*

1973
Coal 24.8%
Oil 45.0%
Other** 0.1%
Combustible Renewables & Waste 11.2%
Hydro 1.8%
Nuclear 0.9%
Natural Gas 16.2%

2002
Coal 23.5%
Oil 34.9%
Other** 0.5%
Combustible Renewables & Waste 10.9%
Hydro 2.2%
Nuclear 6.8%
Natural Gas 21.2%

* Excludes international marine bunkers and electricity trade.
** Other includes geothermal, solar, wind, heat, etc.

Source: Key World Energy Statistics 2004, International Energy Agency

Energy production and supply

Energy – its impact on the environment and society

Introduction

Energy is essential for the smooth running of everyday life and contributes to the nation's wealth and prosperity. The energy industries' output of around £30 billion in 2003 contributed 3.3 per cent of GDP and the industries employed more than 164,000 people. Minimising the environmental impact of the production, distribution and use of fuels is a big challenge.

Oil

Oil supplies 35 per cent of the primary energy used in the UK but can cause considerable harm if released into the environment. It spreads rapidly across water surfaces where it can damage birds, other wildlife and when it reaches shoreline can damage holiday beaches and shellfisheries as well. Oil leakages can occur during extraction, transportation by tankers or pipelines, during refinery activities and during distribution at petrol stations. The size and potential impact of leakages is greatest when crude oil is being transported by sea. The EU imposed a ban on the use of single-hulled tankers for the transport of heavy oils in 2003, which will be phased in by 2010, though the largest tankers will be banned from 2005. Other environmental impacts can occur when drilling wells for exploration or extraction.

Produced water is water that occurs naturally in oil reservoirs that is extracted with oil and gas and then returned into the sea. The average content of oil in produced water for 2003, for the UK Continental Shelf (UKCS) as a whole, was 21 parts per million, slightly lower than in 2000 which was the lowest value yet recorded in the UKCS.

The total amount of oil spilled during 2001 was 94 tonnes. There was also an increase in the number of installations reporting spills (139 in 2001 compared with 117 in 2000). However, of the 436 reports, 96 per cent were of less than 1 tonne.

Gas

Gas has become an increasingly important energy source in the UK in the last few years, particularly its use for electricity generation in place of coal, and comprised 39 per cent of total primary supply in 2003. Thirty per cent of primary gas demand was used in the generation of electricity, including auto-generation.

The energy industries' output of around £30 billion in 2003 contributed 3.3 per cent of GDP and the industries employed more than 164,000 people

---

Gas is extracted from the North Sea at the same point as oil, though it behaves differently to oil as leakages are released into the atmosphere rather than into the sea itself. Flaring is the process used to separate gas from oil, when it is considered too volatile to be transported to land. A small amount of gas is also extracted from onshore sites, including colliery methane from mines. Total carbon dioxide emissions from gas were 56.9 million tonnes in 2003, of which a quarter was due to gas-fired power stations.

Electricity

Although consumption itself has no direct environmental impact, increased consumption requires increased generation with greater overall environmental impact. Efficient use of electricity is therefore an important and effective way to reduce environmental impact of both its generation and supply. Reducing losses in transmission also helps.

The environmental impact of electricity generation is largely dependent on the choice of fuels that are used to generate it. Fossil fuels produce greenhouse gas emissions when combusted as well as other emissions that impact on local air and water quality. Electricity generation from nuclear fission results in radioactive discharges into the air and water.

Between 1990 and 2002, the amount of electricity generated increased by 16.9 per cent. Over the same period, emissions from power stations of sulphur dioxide reduced by 75 per cent, nitrogen oxide fell by 51.4 per cent, and PM10 emissions by 86.1 per cent.

Carbon dioxide emissions from power stations between 1990 and 2003 fell by $15^1/_2$ per cent. Savings due to increased efficiency and fuel switching led to CO_2 emissions being $31^1/_2$ per cent lower in 2003 compared with what they would have been (taking into account increased electricity demand since 1990). About one-half of this was due to changes in the mix of fuels used in the power stations (a

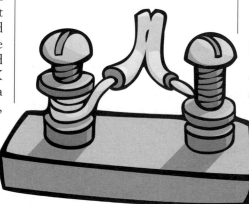

A *future* based on 100% renewable energy

Have you always thought that there would never be enough energy generated from renewable energy? Jackie Carpenter of Energy 21 helps to change your thinking

Alternative energies – wrong ideas

I think that sustainable energy is the most important issue of our time. Fossil and nuclear fuels do not offer long-term security and cause a range of environmental problems. Renewable energy is not an alternative. The planet has run on renewable energy for millions of years, on fossil fuel for only a few hundred and on nuclear for only a few decades. Fossil and nuclear are the alternatives and short-lived alternatives at that. They will be seen by future generations as nothing more than a blip on the timeline.

Global warming

Global warming is a fact. The temperature of the Earth's atmosphere is going up because of the increased concentration of carbon dioxide from people burning fossil fuels. Global warming threatens us with huge and chaotic weather changes and because of possible positive feed-back systems, the whole thing could run out of control even to the extent of the planet becoming uninhabitable for most species. The precautionary principle says don't risk it. We just don't know what will happen.

The reversal of the Gulf Stream is even a possibility. This current of warm water brings a huge amount of heat across the Atlantic from South America to the UK. As global warming melts the ice at the North Pole, new currents of cold water are beginning to flow south. Some observers think this might lead to the reversal of the Gulf Stream. This would not happen slowly; if it happened at all it would flip over-night, leading to huge changes to the British climate. We could then have winters as cold as those of Newfoundland which is on roughly

By Jackie Carpenter BSc CEng MIMechE FRSA

the same latitude and has six months of winter.

Oil is running out!

The next set of problems actually offers a possible solution to the global warming threat. Perhaps it will be impossible to emit enough carbon dioxide to make the planet uninhabitable, simply because we are going to run out of fossil fuels long before then. But running out of fossil fuels is a huge threat in itself.

Our global society is dependent on oil. Oil is the basis of almost everything we do – growing food, making goods, health-care products, transport locally and world-wide. Ever since we found the first drop of oil, we have known that oil is a finite resource. As Dr Colin Campbell has said, 'Few would deny that the world runs on oil. By describing oil as a fossil fuel, everyone admits that it was formed in the past, which means that we started running out when we consumed the first barrel. That much can surely be agreed, even if opinions differ about how far along the depletion curve we are.'

Whether the time-scale for this crisis is five years, ten years or fifty years does not change the argument that sometime during this century we shall have to face up to the biggest and most important challenge that civilisation has ever faced. It would surely be sensible to start working on a new economy independent of oil as soon as possible.

The only safe and sustainable solution is renewable energy!

Our world economy depends on energy for electricity, heat and transport. The fossil and nuclear systems are full of problems including global warming, nuclear pollution and the eventual depletion of both fossil and uranium reserves. The only safe and sustainable long-term solution is renewable energy. Energy efficiency is important but without the change to renewable energy, more efficient use of energy just postpones and prolongs the problems.

What is renewable energy?

Renewable energy comes from sources that are continually replenished by energy from the sun (sometimes by the moon or by heat from within the earth). The energy is flowing past us all the time, and we already rely on it for most of our light and heat. Compare the glorious light of the day with our feeble light bulbs at night and it is easy to see what a small proportion of the energy we need comes from our human devices, compared with that from God.

We can capture the energy of the sun directly through the design of our buildings, solar hot water and PV (photoelectric) cells. The sun evaporates water, which then falls as rain – the source of energy in hydropower. The sun heats the globe differently in different parts, and this causes the air to flow – the source of energy in the wind and waves. Energy from the sun is captured by plants using photosynthesis – the source of energy in food, biomass and plant oils. The gravitational pull between the sun, moon and earth gives rise to the energy in the tides.

The main essence of renewable energy is its diversity. We are not talking about replacing a few large coal-fired power stations with a few massive wind warms. The idea is that the old top-down system of distributing all our energy, whether it be electricity or petrol, is replaced by a plethora of local technologies. Renewable energy is like food (in fact, food is a kind of renewable energy). We don't despair because we cannot grow a cabbage that is 100 miles in diameter; we just grow lots of little cabbages. We don't eat tomatoes for breakfast, dinner and tea every day of the year; we depend on a wide range of different types of food, so that we never run out. So it is with renewable energy. If the wind doesn't blow, it has probably been raining and the rivers will produce hydropower. If the sun doesn't shine, we can use the energy stored by plants in wood or oil.

Can we really power the whole world on renewable energy?

There is plenty of energy from the sun: 6,000 times as much energy as we use, and it will last for as long as the sun shines. There are no significant technical difficulties. We have the wealth and we have the time – it wouldn't take long to build millions of simple, local power plants. Good technology means appropriate technology, often simple and not high-tech. My home runs completely on renewable energy (with a central heating system fuelled by waste wood, solar hot water and a solar electric roof on my log-shed in the garden). The real difficulty in moving towards a renewable energy future is the psychological one of moving away from an old system to a completely new one.

Renewable energy collection can be clean and green. A resource is renewable if it is constantly replenished by energy from outside the Earth. An energy system is

Types of renewable energy

Biofuels
Plant oil e.g. sunflower, rape, peanut. Can make biodiesel. Can use waste food oil. Ethanol e.g. distilled from grain or potatoes.

Biomass
Can be obtained from forestry, food or crop waste or special energy crops. Can be used in wood stoves, boilers or power pant.

Fuel cells
Renewable if the fuel runs on hydrogen produced from water using renewable energy. Part of a hydrogen economy.

Sun-tubes
Perfect for office lighting when the sun is shining. Use light as light.

Passive solar
Use of a building to collect heat from the sun e.g. by adding a conservatory. Good building design means little added heat is needed.

PV – Solar electricity (Photovoltaics)
Light shining on silicon material produces an electric current with no moving parts. Can be used as roofing or cladding.

Solar hot water
Simple use of heat from the sun. Flat plates or vacuum tubes.

Water
Hydropower, tide, wave and undersea current turbines all use power from water. Large hydro is well established. Small hydro is possible at mill sites.

Earth energy
Obtained from hot rocks deep underground or warmth from the oil using ground loops. Heat pumps work like fridges pumping the heat from cold to warm.

Wind
Wind turbines can be owned by the community. Off-shore wind farms developing in the UK.

sustainable if we use the energy at the rate at which it is replenished or less, and if we ensure that anything emitted from the system is non-damaging. Thus wood-fuelled systems are sustainable if we use wood from forests what are continually being replanted, and if we deal sensibly with the smoke and ash.

Renewable energy is available to all people in all parts of the word in various local form, so it is equitable and just. All our energy can be collected locally, wherever we live. In Denmark we collected rape-seeds from the field, minced and filtered them, and then tipped the oil straight into the fuel tank of our plant oil car. There are often no fuel costs (no one can tax the sun or the wind). Renewable energy offers real opportunities for communities to benefit. Local people can reap

financial rewards and feel that they have power over their own future.

Diverse, local solutions are the real challenge as we move forwards to a renewable energy future.

■ Jackie Carpenter was President of the Women's Engineering Society from 2003 to 2003. In 1993 Jackie and her partner, Ian, set up Hebe, a renewable energy consultancy. Later she helped found the charity Energy 21, whose mission is 'to unite action for renewable energy'. She is now the Managing Director of Energy 21. The Energy Store, Estate Yard, Castle Combe, Wiltshire, SN14 7HU. E-mail: info@energy21.org Tel: 01249 783415.

■ The above article first appeared in *Green Christian* – the magazine of the Christian Ecology Link.

© Jackie Carpenter, Energy 21

Has the hydroelectric potential of the UK now largely been exploited?
Economic and environmentally-sensitive considerations mean there is very little scope for more large-scale hydroelectric schemes. Potential new schemes in Scotland have been vigorously opposed. However, small-scale and off-grid micro hydroelectric schemes offer some potential.

Is fuel cell technology only being developed for transport?
The major application at present is in powering vehicles. However, much research is under way, with some combined heat and power schemes in operation. Large stationary hydrogen fuel cells could store power in the future.

How much power are we harnessing from the ocean?
Not as much as the potential suggests. Currently it amounts to 0.5 MW or 0.006% of national electricity consumption, with one scheme, the Islay wave generator. However, more research into wave and tidal renewable technology is under way at a number of locations including the Orkney Islands and Blyth in Northumberland.

Is the use of solar energy restricted to domestic situations e.g. for providing hot water?
No, it also has commercial applications and can be used in hotels, leisure

centres etc. in the case of active solar heating. Photovoltaic cells (PV) technology systems in operation in the UK currently tend to meet small power requirements in applications such as phone booths and roadside monitoring systems, but larger-scale applications on buildings connected to local networks where electricity can be 'stored' by the grid are becoming more commonplace.

Aren't most people opposed to wind farms?
There is some localised opposition to some wind farm developments but

---

The development of onshore wind farms is continuing. Offshore wind farm technology is a new area where development is only beginning

---

public surveys consistently demonstrate that a clear majority of between 70 and 80% of the general public are in favour of wind energy. The same surveys have shown that this positive feeling is highest amongst those living near wind farms. Similar numbers do not believe that wind farms spoil the scenery or cause noise nuisance.

Are most wind farms now going to be built offshore?
The development of onshore wind farms is continuing. Offshore wind farm technology is a new area where development is only beginning. At the end of 2003, developments with the potential of 1,172 MW had been approved onshore and 1,183 MW offshore. The offshore wind farms already built are Blyth Offshore (4MW) and North Hoyle (60 MW). Another, at Scroby Sands (60MW), is under construction. Up to an additional 7.5 GW (7,500 MW) of offshore wind is currently being licensed through the second round of offshore leases, administered by the Crown estate. In total up to 8.7 GW of offshore wind is in the pipeline, a figure greater than the current installed capacity of the UK's AGR nuclear power stations.

■ The above information is from the Department of Trade and Industry's website which can be found at www.dti.gov.uk

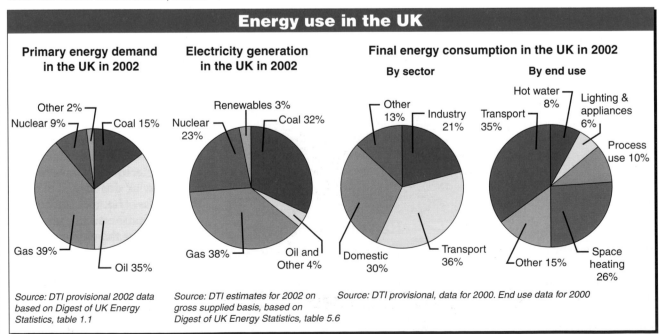

Energy use in the UK

Primary energy demand in the UK in 2002

Other 2%
Nuclear 9%
Coal 15%
Gas 39%
Oil 35%

Source: DTI provisional 2002 data based on Digest of UK Energy Statistics, table 1.1

Electricity generation in the UK in 2002

Renewables 3%
Nuclear 23%
Coal 32%
Gas 38%
Oil and Other 4%

Source: DTI estimates for 2002 on gross supplied basis, based on Digest of UK Energy Statistics, table 5.6

Final energy consumption in the UK in 2002

By sector

Other 13%
Industry 21%
Domestic 30%
Transport 36%

By end use

Hot water 8%
Lighting & appliances 6%
Transport 35%
Process use 10%
Other 15%
Space heating 26%

Source: DTI provisional, data for 2000. End use data for 2000

Lacking energy

**Britain produces comparatively little renewable power –
while the global market for it is growing, writes Mark Tran**

Solar energy is one of the most promising technologies for reducing the greenhouse gases that contribute to global warming, yet Britain is a laggard in developing this clean energy.

Although the UK has invested £25m in solar power – £9m this year – only 10 megawatts of such electricity was produced in 2003, about the same as the output of a small wind farm and only 1% of that produced by the Sizewell nuclear power station, in Suffolk.

Lamenting Britain's slow adoption of solar energy, Peter Hain, the secretary for Wales, argued in July 2004 that every new home should, by law, be fitted with photovoltaic panels to produce solar electricity.

However, Mr Hain's suggestions, like his proposed panels, are unlikely to see the light of day as the government sees limited potential for solar energy in the UK.

A review of innovation in renewable published this year by the department for trade and industry and the Carbon Trust, a group that promotes low carbon use, said: 'Current technology solar PV [photovoltaic] installation is expensive under UK conditions. There may be a future breakthrough in solar PV technologies which could substantially reduce costs, advancing the point at which solar PV is an economical technology under UK conditions. However, the breakthrough and extent of impact are highly uncertain.'

The government's position receives some support from John Mogford, head of BP's gas, power and renewables division. Too diplomatic to say that Mr Hain is full of hot air, he still makes it plain that he thinks the minister is being unrealistic.

'Solar plays different roles in different countries,' he told Guardian Unlimited, in an interview at BP's headquarters in St James' Square in London. 'Even in the US, it's different state by state. Every government should have different policies depending on climatic conditions.'

Mr Mogford is referring, of course, to Britain's lack of sun. But Britain does have plenty of wind. The UK has more wind off its coasts than anywhere else in Europe, and the government believes that windpower holds considerable promise.

Developers have entered into agreements for leasing windfarm sites around the coast with a total capacity of at least 1,400 megawatts of renewable energy, sufficient to power a city the size of Greater Manchester.

Solar energy is one of the most promising technologies for reducing the greenhouse gases

---

Whichever source the government favours, the UK produces less renewable electricity – be it sun, wind or hydro – than several other large European countries. In 2000, renewables (excluding large hydro plant and mixed waste incineration) supplied only 1.3% of Britain's electricity. That compared with 16.7% in Denmark, 4% in the Netherlands, 3.2% in Germany and 3.4% in Spain.

Notwithstanding its poor record, the Blair government has set itself the target of producing 10% of its electricity needs from renewable sources by 2010, a goal that many environmental groups think is unattainable at the present pace.

As Mr Mogford explains, a big hurdle for the UK is that it has a highly developed grid system, one of the world's most advanced, and cheap electricity. Mr Mogford compares renewables to mobile phones. If a country had no land lines, it would make sense to go straight to mobile phones and skip the huge investment in fixed lines.

Similarly, it would be advantageous for a country without a highly developed power grid system to sink resources into solar energy. But in the UK, the incremental cost of adding another power station to

Clearly, any plans to build any new nuclear power stations must be accompanied by a strategy for dealing with the long-term storage and disposal of the radioactive waste that they will produce.

Britain must have a solution to the long-term storage and disposal of existing radioactive waste, much of which was a product of the civil and military nuclear programmes of the 1950s, and new waste from the operation and decommissioning of the present generation of nuclear power stations. However, we do not necessarily need to have this solution before making a decision about the building of new nuclear power stations.

If we are to tackle the ominous spectre of climate change, we must seriously consider all the options open to us and – whether we are from the political, scientific or environmental communities – not fight shy of making and defending difficult decisions.

■ Lord May of Oxford is president of the Royal Society, the UK national academy of science, and was chief scientific adviser to the Government 1995-2000.

Nuclear energy

Information from British Nuclear Fuels Plc

The various activities associated with the production of electricity from nuclear reactions are referred to collectively as the nuclear energy cycle.

The nuclear energy cycle starts with the mining of uranium and ends with the disposal of nuclear waste. Fuel removed from a reactor, after it has reached the end of its useful life, can be reprocessed (similar to recycling) to produce new fuel. With this option for nuclear fuel, the activities can now form a complete cycle.

Uranium

Uranium is a slightly radioactive metal that occurs throughout the earth's crust. It is about 500 times more abundant than gold and about as common as tin. It is present in most rocks and soils, as well as in many rivers and in seawater. It occurs, for example, in concentrations of about four parts per million (ppm) in granite, which makes up 60% of the earth's crust. In fertilisers, uranium concentration can be as high as 400 ppm (0.04%). And some coal deposits contain uranium at concentrations greater than 100 ppm (0.01%). There are a number of areas around the world where uranium is extracted for use as nuclear fuel. In these places, the concentration of uranium in the ground is sufficiently high to make it economically feasible.

Uranium mining

The characteristics of a uranium deposit – such as how far underground it is – will determine how it is mined. Deep deposits are usually mined

through underground shafts and shallower deposits by open-pit methods. Uranium also is produced as the by-product of other mining operations, like gold, copper and phosphates. A newer way to mine uranium is leaching the uranium out of the ground. This method is popular because of its economic and environmental advantages.

Uranium milling

To concentrate the uranium content, the ore is ground, treated and purified using chemical and physical processes. The result is a solid uranium ore concentrate, which contains more than 60% uranium. This is commonly referred to as 'yellowcake'. The concentrate has a smaller volume than the ore and hence is less expensive to ship.

Conversion

The uranium ore is converted into uranium hexafluoride, which can be enriched to produce fuel for most types of reactors. Alternatively, the uranium ore can be processed to produce uranium metal. This can be used as the fuel for the UK's Magnox reactors, which do not require enriched uranium.

Enrichment

Natural uranium is composed of two types or isotopes, namely U-235 and U-238. Only U-235 is capable of sustaining fission (or atom splitting). This is what creates the energy to run a nuclear power plant. However, U-235 makes up less than 1% of natural uranium. For uranium to be usable as nuclear fuel, a higher concentration of U-235 is required. The enrichment process produces this higher concentration, typically between 3.5% and 4.5% U-235, by removing a large part of the U-238 (80% for enrichment to 3.5%).

There are two enrichment processes in large-scale commercial use, each of which uses uranium hexafluoride as feed. They are gaseous diffusion and gas centrifugation. The product of this stage of the nuclear fuel cycle is enriched uranium hexafluoride. This is reconverted to produce enriched uranium oxide.

Power generation

The generation process starts with the splitting of uranium atoms in a controlled way inside a reactor. This produces heat energy. An atom consists of protons, neutrons and electrons. When the atom is split, two or three neutrons are thrown off at tremendous speed – about 10,000 miles per second. The reactor's graphite core or moderator slows down these neutrons. This enables them to hit other uranium atoms, resulting in another split. More heat energy and more neutrons are released, creating a 'chain reaction'.

The fissioning of uranium is used as a source of heat in a nuclear power station. It works in the same way that the burning of coal, gas or oil is used as a source of heat in a thermal power plant. The core contains vertical channels for uranium fuel elements and control rods. These rods are made of boron steel. They absorb neutrons to allow the reactor to be started or shut down and to control its power level. Either carbon dioxide or water is pumped through the channels. This cools the fuel and transfers the heat energy to boilers to produce steam. The steam is used to drive a turbine connected to a generator, which produces electricity.

Spent fuel

Over time, the concentration of fission fragments in a fuel bundle increases. Eventually, it reaches the point where it is no longer practical to continue to use the fuel. At this time, the 'spent fuel' is removed from the reactor. The energy produced from a fuel bundle varies with the type of reactor and the policy of its operator.

Typically, more than 40 million kilowatt-hours of electricity are produced from one tonne of natural uranium. To produce this amount of electrical power from fossil fuels would require the burning of over 16,000 tonnes of black coal or 80,000 barrels of oil.

Spent fuel storage

When removed from a reactor, a fuel bundle will be emitting radiation – primarily from the fission fragments – and heat. Spent fuel is unloaded into a storage facility immediately adjacent to the reactor. This is to allow the radiation levels and the quantity of heat being released to decrease.

These facilities are large pools of water. The water acts as a shield against the radiation and as an absorber of the heat released. Spent fuel is generally held in such pools for a minimum of about five months.

Ultimately, spent fuel must either be reprocessed or sent for permanent disposal.

Reprocessing

Spent fuel is about 95% U-238. But it also contains U-235 that has not fissioned, plutonium and fission products, which are highly radioactive. In a reprocessing facility, the spent fuel is separated into its three components: uranium, plutonium and waste containing fission products. Reprocessing facilitates recycling and produces a significantly reduced volume of waste.

Uranium and plutonium recycling

The uranium from reprocessing typically contains a slightly higher concentration of U-235 than occurs in nature. It can be reused as fuel after conversion and enrichment, if necessary. The plutonium can be made into Mixed Oxide (MOX) fuel, in which uranium and plutonium oxides are combined.

In reactors that use MOX fuel, plutonium substitutes for U-235 as the material that fissions and produces heat for steam production and neutrons to sustain a chain reaction.

Spent fuel disposal

What happens to spent fuel not destined for reprocessing and the waste from reprocessing? Currently, there are no disposal (as opposed to storage) facilities into which such material can be placed. There is a reluctance to dispose of spent fuel because it represents a resource. It can be reprocessed at a later date to allow recycling of the uranium and plutonium.

A number of countries are carrying out studies to determine the optimum approach to the disposal of spent fuel and waste from reprocessing. The most commonly favoured method for disposal under consideration is placement into deep geological formations. This would involve cooling the spent fuel, probably in dry stores above ground, for several years. Then it would be conditioned, packed and buried in a deep repository.

Waste

Wastes from the nuclear fuel cycle are categorised as high-, medium- or low-level. This is according to the amount of radiation that they emit. Such wastes come from a number of sources and include:

- essentially non-radioactive waste resulting from mining
- low-level waste produced at all stages of the fuel cycle
- intermediate-level waste produced during reactor operation and by reprocessing
- high-level waste, which is spent fuel and waste containing fission products from reprocessing.

The enrichment process leads to the production of 'depleted' uranium. This is uranium in which the concentration of U-235 is significantly less than the 0.7% found in nature. Small quantities of this material, which is primarily U-238, are used in applications where high density material is required. This includes radiation shielding and the production of MOX. While U-238 is not fissionable it is a low specific activity radioactive material. Therefore some precautions must be taken in its storage or disposal.

Clean-up

Clean-up of nuclear weapons material

Nuclear weapons built years ago are being dismantled. Their uranium is being retrieved for use as fuel in nuclear power plants. Nuclear weapons contain highly enriched uranium, whose U-235 content is 20-90%. The Russian and US Governments are working with the nuclear energy industry to bring the enrichment level below 5%. In this way, old dismantled weapons will leave a peaceful legacy. They will provide electricity for thousands of daily uses.

- The above information is from British Nuclear Fuels Plc's website which can be found at www.bnfl.com

© *British Nuclear Fuels Plc*

KEY FACTS

■ Burning coal, oil and gas inevitably produces carbon dioxide gas, a gas which plays a major role in climate change. (p. 1)

■ The easiest way to reduce reliance on fossil fuels is to generate and use energy more efficiently. (p. 2)

■ Coal and other solid fuels contributed 16.6 per cent of UK primary energy supply in 2003, over half being imported mainly from South Africa, Australia and Russia. (p. 4)

■ Think of the electricity you use in a day. You are woken by the clock radio buzzing into life, and you turn the bathroom light on as you climb into your power shower. After dressing you head downstairs, where you turn on another radio, put some bread into the toaster and turn on the kettle, getting the milk from your fridge to put in your tea. After breakfast you head to work, where the lights are burning – and on go the computer and desktop fan. (p. 6)

■ Renewable energy comes from sources that are continually replenished by energy from the sun (sometimes by the moon or by heat from within the earth). (p. 8)

■ Renewable energy can replace oil, gas and coal completely, and power the world without the emissions that cause global warming. (p. 10)

■ Solar energy is one of the most promising technologies for reducing the greenhouse gases. (p. 11)

■ The development of onshore wind farms is continuing. Offshore wind farm technology is a new area where development is only beginning. (P. 12)

■ Lamenting Britain's slow adoption of solar energy, Peter Hain, the secretary for Wales, argued in July 2004 that every new home should, by law, be fitted with photovoltaic panels to produce solar electricity. (p. 13)

■ Hydro installations can have a useful life of over 100 years – many such plants are in existence worldwide. (p. 14)

■ Government legislation requires that by 2010, 10% of electricity supply must come from renewable sources. Wind power is currently the most cost-effective renewable energy technology in a position to help do that. Around 3,500 additional modern wind turbines are all that would be needed to deliver 8% of the UK's electricity by 2010, roughly 2,000 onshore and 1,500 offshore. (p. 15)

■ Biomass was the first fuel that mankind learned to use for energy; the first fires of primitive man burning wood for warmth and cooking. (p. 18)

■ Nuclear power stations currently produce about a quarter of Britain's electricity. Many are now too old to continue to operate efficiently and safely and are being closed down. (p. 20)

■ The UK's emissions of greenhouse gases fell substantially from 1990 onwards, mainly because of the switch from coal to gas – a fuel which produces less CO_2. However, if matters were left to themselves this downwards trend would not continue. (p. 24)

■ Did you know that both wind and wave power could meet our electricity needs three times over? That solar power alone could produce two-thirds of it? That wind parks off the coast of East Anglia alone could replace all of our nuclear power stations? (p. 26)

■ Around 33% of the heat lost in your home is through the walls, so insulating them can be the most cost-effective way to save energy in the home. (p. 30)

■ Remember the three Rs – re-use, repair, recycle! They're more beneficial in that order – it's better to find another use for something or to use it again; if it is broken, repair it; and if you can't do either, take it to be recycled. Anything is better than landfill! (p. 32)

■ £1.2 billion worth of electricity a year goes on cooling and freezing food and drinks in the UK. (p. 32)

■ In the UK, energy experts suggest we use £800 million worth of electricity using washing machines, tumble dryers and dishwashers. This alone produces 5 million tonnes of carbon dioxide each year. (p. 33)

■ Two in five people in England claimed to have cut down their use of electricity and gas in 2001, according to the DEFRA survey, and 80 per cent of those who had made cuts had done so to save money. Only 15 per cent, equivalent to 6 in every 100 of all respondents to the survey, mentioned environmental concerns as a motivation to cut their energy consumption. (p. 35)

■ The government wants to see 10% of the country's electricity power derived from renewable sources by 2010 and 20% by 2020. But it is difficult to see how those targets will be reached. Britain lags behind in Europe when it comes to renewables. In 2000, renewable sources – excluding large hydroelectric plants and mixed waste incinerators – supplied only 1.3% of the country's electricity, compared with 16.7% in Denmark, 4% in the Netherlands, 3.4% in Spain and 3.2% in Germany. (p. 39)

You might like to contact the following organisations for further information. Due to the increasing cost of postage, many organisations cannot respond to enquiries unless they receive a stamped, addressed envelope.

British BioGen
Rear North Suite, 7th Floor
16 Belgrave Square
London, SW1X 8PQ
Tel: 020 7235 8474
Fax: 020 78317223
E-mail: info@britishbiogen.co.uk
Website: www.britishbiogen.co.uk
Promotes and co-ordinates the commercial development of biomass as a renewable fuel resource for energy production.

British Hydropower Association
Unit 12, Riverside Park
Station Road, Wimborne
Dorset, BH21 1QU
Tel: 01202 886622
Fax: 01202 886609
E-mail: info@british-hydro.org
Website: www.british-hydro.org
Protects and promotes the interests of water-power producers, users and organisations who service the needs of the industry.

British Nuclear Fuels Plc
Hinton House, Risley
Warrington, WA3 6AS
Tel: 01925 832000
Website: www.bnfl.com
BNFL provides high quality, cost-effective nuclear fuel cycle services to customers.

British Wind Energy Association (BWEA)
Renewable Energy House
1 Aztec Row, Berners Road
London, N1 0PW
Tel: 020 7689 1960
Fax: 020 7402 7107
E-mail: info@bwea.com
Website: www.bwea.com
The trade and professional body for the UK wind industry.

Campaign to Protect Rural England (CPRE)
128 Southwark Street
London, SE1 0SW
Tel: 020 7981 2800
Fax: 020 7981 2899
E-mail: info@cpre.org.uk
Website: www.cpre.org.uk

CPRE exists to promote the beauty, tranquillity and diversity of rural England by encouraging the sustainable use of land and other natural resources in town and country.

Centre for Alternative Technology (CAT)
Machynlleth
Powys, SY20 9AZ
Tel: 01654 702400
Fax: 01654 702782
E-mail: info@cat.org.uk
Website: www.cat.org.uk
A display and education centre. Its seven-acre site has working displays of wind, water and solar power, low energy building, organic growing and alternative sewage systems.

Energy Saving Trust
21 Dartmouth Street
London, SW1H 9BP
Tel: 020 7222 0101
Fax: 020 7654 2444
Website: www.est.org.uk
One of the UK's leading organisations addressing the damaging effects of climate change.

Environ
Parkfield, Hinkley Road
Western Park
Leicester, LE3 6HX
Tel: 0116 222 0222
Fax: 0116 255 2343
E-mail: enquiries@environ.org.uk
Website: www.environ.org.uk
Environ is an independent charity working to improve the environment and the communities we live in.

Friends of the Earth (FOE)
26-28 Underwood Street
London, N1 7JQ
Tel: 020 7490 1555
Fax: 020 7490 0881
E-mail: info@foe.co.uk
Website: www.foe.co.uk
Inspires solutions to environmental problems which make life better for people.

Greenpeace
Canonbury Villas
London, N1 2PN
Tel: 020 7865 8100
Fax: 020 7865 8200
E-mail: gn-info@uk.greenpeace.org
Website: www.greenpeace.org.uk
Greenpeace is an independent non-profit global campaigning organisation that uses non-violent, creative confrontation to expose global environmental problems and their causes.

International Energy Agency (IEA)
9 rue de la Fédération
75739 Paris Cedex 15 , France
Tel: + 33 1 40 57 65 51
Fax: +33 1 40 57 65 59
E-mail: info@iea.org
Website: www.iea.org
An autonomous agency linked with the Organisation for Economic Co-operation and Development (OECD).

Network for Alternative Technology and Technology Assessment (NATTA)
c/o Energy and Environment Research Unit
Open University, Walton Hall
Milton Keynes, MK7 6AA
Tel: 01908 654638
Fax: 01908 858407
Website: eeru.open.ac.uk/natta/rol.html
NATTA is an independent information service which focuses on sustainable energy developments and associated issues.

The National Energy Foundation
Davy Avenue, Knowlhill
Milton Keynes, MK5 8NG
Tel: 01908 665555
Fax: 01908 665577
E-mail: nef@natenergy.org.uk
Website: www.natenergy.org.uk
Works for the more efficient, innovative, and safe use of energy and to increase the public awareness of energy in all its aspects.

INDEX

ACKNOWLEDGEMENTS

The publisher is grateful for permission to reproduce the following material.

While every care has been taken to trace and acknowledge copyright, the publisher tenders its apology for any accidental infringement or where copyright has proved untraceable. The publisher would be pleased to come to a suitable arrangement in any such case with the rightful owner.

Chapter One: Energy Alternatives

The case for renewable energy, © Network for Alternative Technology and Technology Assessment (NATTA), *Fuel shares of total primary energy supply*, © International Energy Agency, *Energy production and supply*, © Crown copyright is reproduced with the permission of Her Majesty's Stationery Office, *Renewables explained*, © Crown copyright is reproduced with the permission of Her Majesty's Stationery Office, *The balance of power*, © Guardian Newspapers Limited 2004, *Producers of nuclear electricity*, © International Energy Agency, *A future based on 100% renewable energy*, © Jackie Carpenter, Energy 21, *Are we ready for when the oil runs out?*, © Guardian Newspapers Limited 2004, *Renewable energy*, © Crown copyright is reproduced with the permission of Her Majesty's Stationery Office, *Energy use in the UK*, © Crown copyright is reproduced with the permission of Her Majesty's Stationery Office, *Lacking energy*, © Guardian Newspapers Limited 2004, *10 things you should know*, © British Hydropower Association, *Wind energy*, © The British Wind Energy Association, *Bioenergy*, © British BioGen, *Nuclear power*, © Friends of the Earth, *The nuclear option*, © Telegraph Group Limited, London 2004, *Nuclear energy*, © British Nuclear Fuels Plc, *Energy*, © Campaign to Protect Rural England, *Renewable energy: the way forward*, © Greenpeace, *Renewable energy utilisation 2003*, © Crown copyright is reproduced with the permission of Her Majesty's Stationery Office.

Chapter Two: Energy Efficiency

Urban myths or simple truths?, © National Energy Foundation, *10 point energy plan*, © Energy Saving Trust 2004, *How green are you?*, © The Centre for Alternative Technology Charity Limited, *Home energy use*, © Energy Saving Trust, *Save energy around the home*, © Environ, *Public attitudes to energy and the environment*, © Crown copyright is reproduced with the permission of Her Majesty's Stationery Office, *Knowledge of major factors contributing to climate change*, © Crown copyright is reproduced with the permission of Her Majesty's Stationery Office, *In winter . . .*, © Powergen Energy Monitor 2003, *Main reasons for taking energy saving steps*, © Powergen Energy Monitor 2003, *Stamp duty cut urged for energy-saving homeowners*, © Telegraph Group Limited, London 2004, *Electricity from renewable sources*, © The Centre for Alternative Technology Charity Limited, *Paid as you burn*, © Guardian Newspapers Limited 2004.

Photographs and illustrations:

Pages 1, 15, 31: Simon Kneebone; pages 10, 28: Angelo Madrid; pages 11, 24: Don Hatcher; page 13: Bev Aisbett; page 37: Pumpkin House.

Craig Donnellan
Cambridge
January, 2005